T0160567

How to Make a Metal Machine

R.D. Mann

Detailed construction including plans, operating instructions, how to build a
wheel and a plan of an ox-cart with wood block bearings showing the wheel
and axle design in place

Practical
ACTION
PUBLISHING

Intermediate Technology Publications 1987

Acknowledgements

We are most grateful for the permission given by J. ABBEY Engineering Works at Watton, nr. Thetford, Norfolk, to use the pivot principle design of their Norfolk Universal Bender on which this metal-bending machine has been based.

We also acknowledge the helpful assistance given by staff of the National College of Agriculture at Silsoe and the Tanzanian Agricultural Machinery Testing Unit in the design, development and construction of the metal-bending machine.

Sketches and photographs were provided by Bob Mann and TAMTU.

Financial assistance for the re-issue of this manual has been provided by the Overseas Development Administration.

Practical Action Publishing Ltd
25 Albert Street, Rugby, CV21 2SD, Warwickshire, UK
www.practicalactionpublishing.com

© Intermediate Technology Publications 1987

First published 1987\Digitised 2013

ISBN 10: 0 90303 129 9
ISBN 13: 9780903031295
ISBN Library Ebook: 9781780442389
Book DOI: http://dx.doi.org/10.3362/9781780442389

Since 1974, Practical Action Publishing has published and disseminated
books and information in support of international development work
throughout the world. Practical Action Publishing is a trading name
of Practical Action Publishing Ltd (Company Reg. No. 1159018), the
wholly owned publishing company of Practical Action. Practical Action
Publishing trades only in support of its parent charity objectives and any
profits are covenanted back to Practical Action (Charity Reg. No. 247257,
Group VAT Registration No. 880 9924 76).

Contents

Foreword

Research to increase the sum of human knowledge carries with it the double responsibility of recording the new information so obtained and of communicating it to others. Too often the exciting work of extending the frontiers of experience is regarded by the investigator as a complete and worthy end in itself and the concomitant duty to revise and enlarge the chart of established knowledge is neglected.

In no field of endeavour is the obligation to record and publish the results of investigation stronger than in the realm of intermediate technology. This is due in part to the general scarcity of published information in this already neglected field. It is also due to the fact that the number of human beings in the world which will benefit from such publications is vastly greater than that of those who stand to benefit from advances at the more sophisticated levels taken for granted in much of Europe and North America.

It therefore gives very great pleasure to introduce the accompanying report on the development of a metal-bending machine. This is one of the first of a series published by the Intermediate Technology Development Group Ltd. designed to make widely available the results of current progress in its work. At the present time, the Group is engaged in fostering research and development in intermediate technology in many parts of the world. This report is convincing evidence of the determination to publish the results of such work in a manner well suited to the requirements of the many who stand to benefit from it.

WYE COLLEGE   H. S. DARLING

Introduction

The establishment of blacksmithing facilities in rural areas of developing countries is essential wherever local construction and maintenance of equipment for small farms is being encouraged.

This low-cost pivot-principle hand-operated metal-bending machine has been designed so that it can be constructed locally, since it is fabricated of easily available mild steel flat, angle, bar and pipe materials.

The machine's main feature is its ability to form wheel rims from cold flat mild steel up to 4" x 3/8" (101 mm x 9.5 mm) cross section, for use on farm carts and other basic agricultural equipment. It can also be used for bending notched angle iron and flat mild steel to whatever angles are required.

Description

Letters in brackets ( ) refer to drawings and photographs which follow.

The basic machine consists of a fixed arm (A) bolted to two pieces of channel (F), the latter provided with holes at both ends and intended to facilitate mounting of the machine on a bench or other solid surface. The pivoting arm (B) which controls the metal bending operation is fitted with a handle (E), and pivot pin (G) which can be placed in any of the nine position holes to fix the arms about the required point according to the bending work involved.

The notched angle iron bending former (L) and the material grip/former (J) were intentionally designed with sides angled at 87½° to allow for material 'spring-back' when bending angle or flat to 90°.

The circle bending former (H) was originally built with a bending surface curve of 14" (355 mm) radius, and when bending 4" x 3/8" black mild steel flat this gave a hoop of 33" nominal diameter. Since the aim was to produce a wheel rim of 30" nominal diameter, being that used in a standard ox-cart design (S), a new circle former of 12½" (317 mm) radius was built; bending trials were carried out, and after the addition by tack welding of two pieces of 4" x 3/8" (101 mm x 9.5 mm) the then total radius of 13¼" (336 mm) was found to give a circle of the required 30" nominal diameter. During bending trials it was found that a flat section of up to 3½" in length remained at both ends of the wheel rim and that there was a small amount of stretching of the material during bending; these points are dealt with in the section on 'Building a Wheel'.

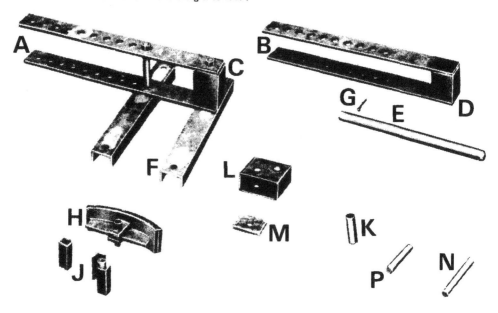

A	Fixed arm	H	Circle bending former
B	Pivoting arm	J	Material grip/former
C	Fixed arm box	K	Cylindrical roller/former
D	Pivoting arm box	L	Angle-iron bending former
E	Handle	M	Distance piece/material guide
F	Mounting supports	N	Fixed arm fittings pin
G	Pivot pin	P	Pivoting arm fittings pin

Construction

List of Parts — Basic Machine

Part	Name	Quantity	Dimensions (mm)
A	Fixed arm	2	12.5 x 76 x 762 M.S. flat
C	Fixed arm box	2	135 x 76 x 76 M.S. angle
F	Mounting supports	2	76 x 38 M.S. channel (508 long)
B	Pivoting arm	2	12.5 x 76 x 762 M.S. flat
D	Pivoting arm box	2	105 x 76 M.S. angle
E	Handle	1	35 Dia. M.S. bar (508 long)
G	Pivot pin	1	25 Dia. M.S. bar (229 long)
	Fixed arm bolts	2	22 Dia. (190-200 long shanks)
	Mounting support bolts	4	20 Dia. (length to suit mounting deck)

List of Parts — Machine Fittings and Formers

Part	Name	Quantity	Dimensions (mm)
H	Circle bending former	1	16 x 101 x 340 curved to 336 radius
		1	12.5 x 76 x 279 M.S. flat
		2	12.5 x 51 x 101 M.S. flat
		1	25 nominal bore pipe (101 long)
J	Material grip/former	2	101 x 38 x 38 M.S. angle
		1	25 nominal bore pipe (101 long)
K	Cylindrical roller/former	1	25 nominal bore pipe (101 long)
L	Angle-iron bending former	1	76 x 152 M.S. box section (140 long)
M	Distance piece/material guide	1	12.5 x 76 x 101 M.S. flat
N	Fixed arm fittings pin	1	25 Dia. M.S. bar (178 long)
P	Pivoting arm fittings pin	1	25 Dia. M.S. bar (140 long)

Note:

1. All parts of the machine and its fittings/formers are made of ordinary mild steel of carbon content no higher than that of En 1A Carbon steel.

2. The circle bending former (H) is illustrated exactly as it appeared after modification to increase its radius to that which would provide a 30" (762 mm) nominal diameter rim from 101 mm x 9.5 mm cross section flat black mild steel. The outer curved surface of this former can be made of single piece of 16 mm x 101 mm x 340 mm mild steel flat and will have adequate strength to withstand the bending forces involved.

3. The outer edges of the heads of the fittings pins (NP) can be hammered to splay them over so that they will not fall through the arm holes.

4. The 25 nominal bore pipe used in fittings and formers is ordinary heavy gauge galvanised steel water pipe.

5. If on initial assembly of the machine it is found that the pins (GNP) are not an easy slide fit, the pivot pin holes and the 25 mm nominal bore pipe pieces can all be lightly reamed or filed to provide the slight clearance required.

Basic Machine Parts

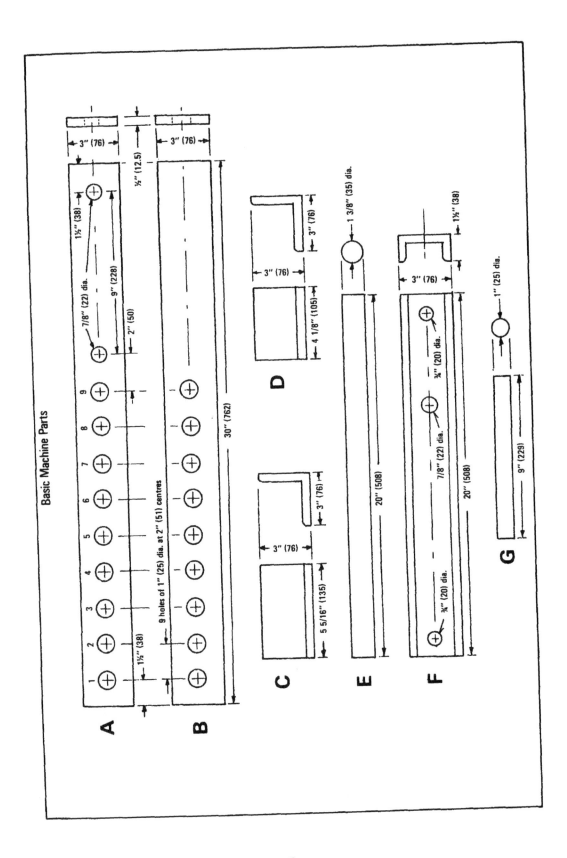

7

Machine Fittings and Formers for Metal Bending

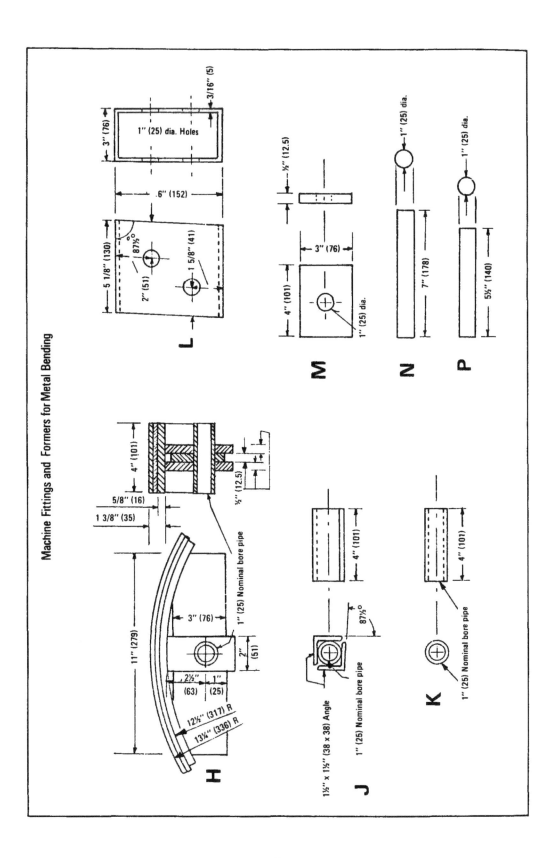

3/16" (5)

3" (76)

1" (25) dia. Holes

.6" (152)

87½°

5 1/8" (130)

2" (51)

1 5/8" (41)

L

½" (12.5)

3" (76)

4" (101)

1" (25) dia.

M

1" (25) dia.

7" (178)

N

1" (25) dia.

5½" (140)

P

4" (101)

5/8" (16)

1 3/8" (35)

½" (12.5)

1" (25) Nominal bore pipe

11" (279)

3" (76)

2" (51)

2½" (63)

1" (25)

12½" (317) R

13¼" (336) R

H

4" (101)

87½°

1½" x 1½" (38 x 38) Angle

1" (25) Nominal bore pipe

J

4" (101)

1" (25) Nominal bore pipe

K

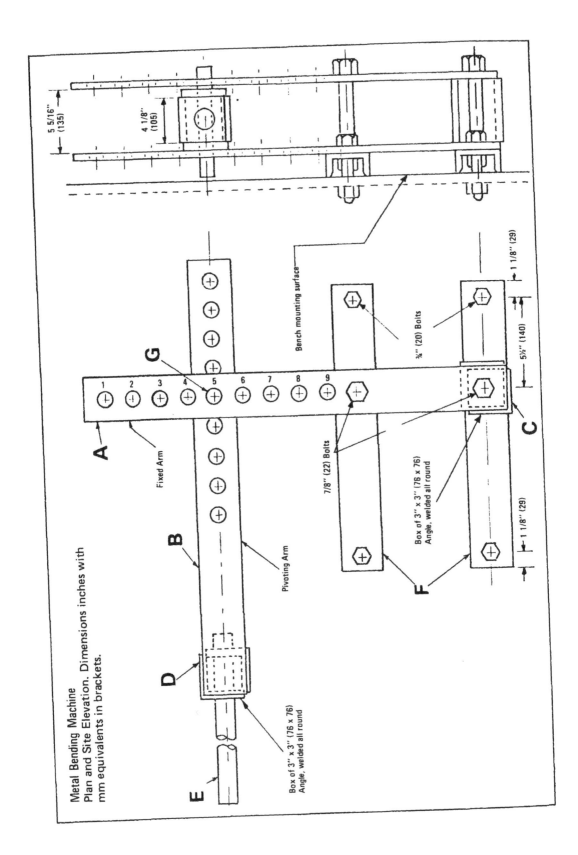

Metal Bending Machine
Plan and Site Elevation. Dimensions inches with mm equivalents in brackets.

5 5/16" (135)

4 1/8" (105)

Bench mounting surface

¾" (20) Bolts

1 1/8" (29)

5½" (140)

1 1/8" (29)

7/8" (22) Bolts

Box of 3" x 3" (76 x 76)
Angle, welded all round

A

Fixed Arm

G

1 2 3 4 5 6 7 8 9

B

Pivoting Arm

D

Box of 3" x 3" (76 x 76)
Angle, welded all round

E

C

F

Assembly

1.   To construct the fixed arm (A1), two pieces of 12.5 mm x 76 mm x 762 mm mild steel flat are clamped one on top of the other, and the nine 25 mm diameter pivot pin holes and two 22 mm diameter bolt holes are drilled through both pieces. Two pieces of angle (C) forming the box are placed in position between the drilled flats (A), 25 mm diameter bars placed through pivot pin holes No. 1 and No. 9; a 135 mm distance piece placed at the outer end and the box-end 22 mm diameter bolt tightened to ensure precise alignment, following which the angles (C) are welded all round and to the flats (A) to form the box.

2.   In building the pivoting arm (B2), the two pieces of mild steel flat (B) are clamped together and the nine 25 mm diameter pivot pin holes drilled as described for the fixed arm. Two pieces of angle (D) are then clamped in box position and welded along their joining edges, followed by drilling through the angle-iron box a hole of 35 mm diameter to take the handle (E). The angle-iron box is then placed in position between the drilled flats (B), a 105 mm distance piece being inserted at the other end and 25 mm diameter bars placed through pivot pin holes No. 1 and No. 9 to obtain correct alignment, the angle-box then being welded all round to the flats (B). One end of the handle (E2) is drilled with a 5 mm diameter hole approx. 15 mm distance from the handle end to take a retaining pin.

3.   The two machine mounting supports (F1) are each drilled with one 22 mm diameter hole and two 20 mm diameter holes, the latter to take the mounting support bolts.

To assemble the basic machine the fixed arm (A) is first bolted securely to the mounting supports (F), the latter being then bolted solidly to the mounting surface deck by 20 mm diameter bolts of the required length. It is important to note that a distance piece 135 mm long, such as a piece of 25 mm bore pipe, should be used for the 2nd fixed arm bolt to pass through so that it can be tightened and still ensure that the fixed arm flats are an equal distance apart throughout their entire length. The assembled basic machine with the pivot pin (G) inserted through both arms is shown in (3). A small weld nodule can be tacked to the side of the top end of the pivot pin to prevent it from falling through the arms or alternatively the bottom of the pivot pin can rest on the mounting deck.

The basic machine

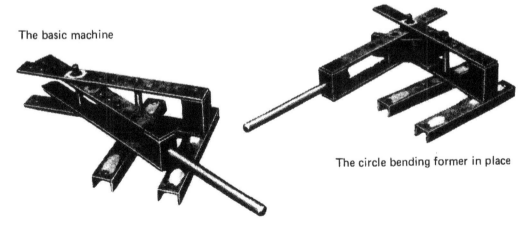

The circle bending former in place

The circle bending former (H) is placed in position, the pivot pin passing through holes No. 5 in the machine arms. The cylindrical roller/former (K) is held in hole No. 8 position in the pivoting arm by fittings pin (P). A material grip/former (J) is placed centrally on top of the distance piece/material guide (M) both being held in hole No. 8 position in the fixed arm by fittings pin (N). The machine set up as above is now ready for the bending operation.

Building a Wheel

From experimentation during prototype bending trials it was found that the following procedure facilitated the forming of a complete circle:

a.   Take a piece of 4" x 3/8" (101 mm x 9.5 mm) black mild steel flat and cut it to 101" (2565 mm) in length.

b.   One operator pushes a 6' (1830 mm) length of 1½" (38 mm) bore pipe over the pivoting arm handle (to provide additional leverage), the pivoting arm is swung back so that the fittings are well clear of the circle bending former and the second operator inserts one end of the mild steel flat in between the fittings and the curved former.

c.   The mild steel flat is held horizontal with its end just within the grip between the material gripper fitting on the fixed arm and the surface of the curved former as the first operator swings the pivoting arm forwards, the forward movement then being continued by pushing with full force on the leverage pipe so that the cylindrical roller on the pivoting arm forces the flat material round until it has curved and is held against the curved former.

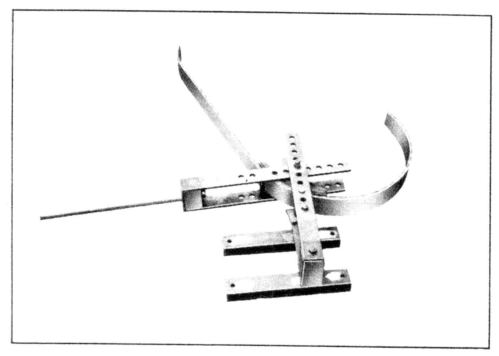

d.   The bending then proceeds by the first operator releasing the grip on the material by drawing the pivoting arm backwards, the second operator pushing the material in past the former by about 2" (50 mm) and the cylindrical roller is then forced hard against the material once more until it is pushed hard against the curved former.

e.   The bending operation continues in this manner by approx. 2" (50 mm) steps until a length of about 12" (305 mm) has been curved. The material is then removed and the same initial bending procedure applied to the other end for a distance of about 12" (305 mm).

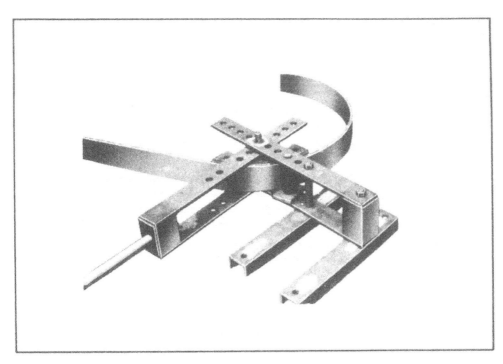

At both extreme ends of the material flat sections will be observed, each of up to 3½''
(89 mm) in length. These flat sections are cut off to leave a total length of material of
94'' (2387 mm).

f.   The material is then re-inserted into the machine and the bending operation continued
     as illustrated. It is of assistance if the top edge of the material is marked off in 2''
     (50 mm) steps by chalk marks to act as a guide each time the material is pushed forward
     for the next bend.

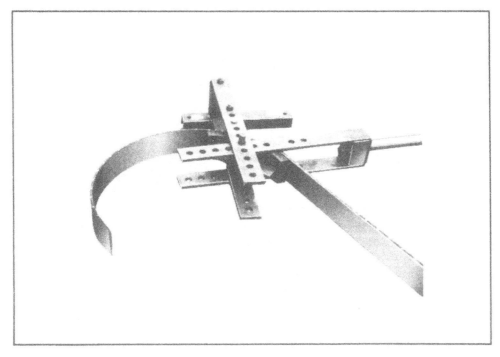

The bending continues until a complete rim has been formed, when it will be found that the curved ends of the material have met on the surface of the curved former. (Note: to avoid distortion of the wheel rim during the bending operation, the material should be held horizontally as it is inserted and moved forward through the machine, and the curved portion of the material should be supported on the same plane as that of the distance piece/material guide (M) as it moves progressively round a curve.)

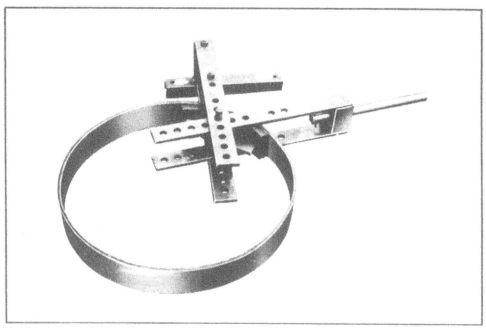

With reference to the given ox-cart design (S), simple jig assembly (QR) was built to facilitate the construction of the wheel and live axle, the procedure being as follows:

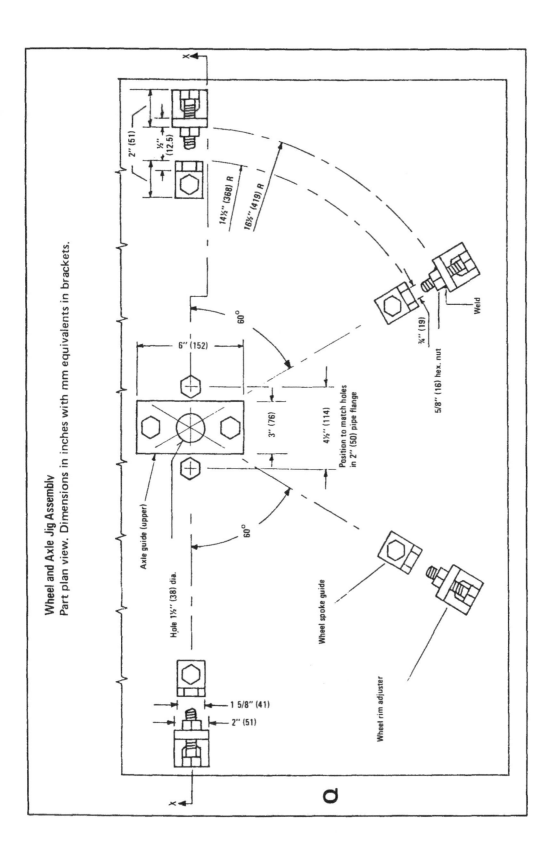

Wheel and Axle Jig Assembly
Part plan view. Dimensions in inches with mm equivalents in brackets.

2" (51)

½" (12.5)

14½" (368) R

16½" (419) R

Weld

6" (152)

¾" (19)

5/8" (16) hex. nut

60°

3" (76)

4½" (114)

Position to match holes
in 2" (50) pipe flange

Axle guide (upper)

60°

Wheel spoke guide

Hole 1½" (38) dia.

Wheel rim adjuster

1 5/8" (41)

2" (51)

14

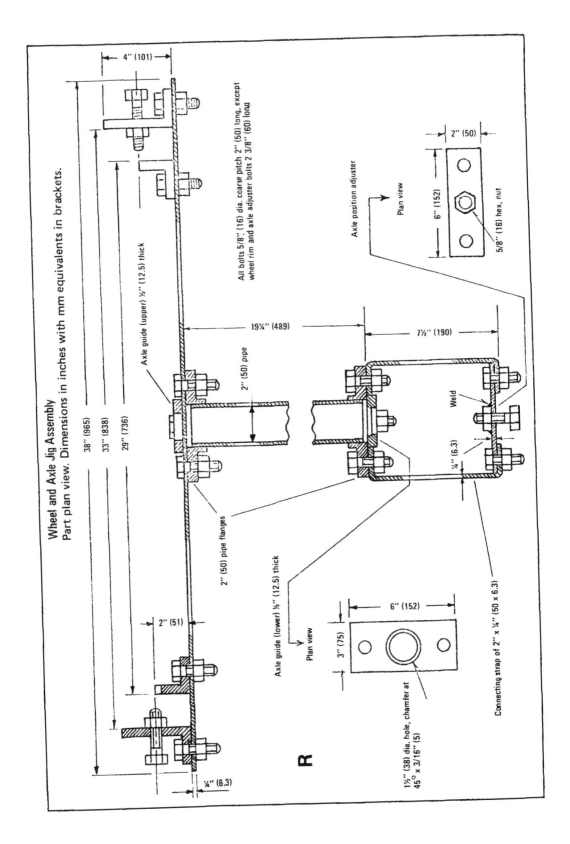

Wheel and Axle Jig Assembly
Part plan view. Dimensions in inches with mm equivalents in brackets.

4" (101)

All bolts 5/8" (16) dia. coarse pitch 2" (50) long, except
wheel rim and axle adjuster bolts 2 3/8" (60) long

Axle position adjuster

2" (50)

Plan view

6" (152)

5/8" (16) hex. nut

Axle guide (upper) ½" (12.5) thick

19¼" (489)

7½" (190)

38" (965)

33" (838)

29" (736)

2" (50) pipe

Weld

¼" (6.3)

2" (50) pipe flanges

Axle guide (lower) ½" (12.5) thick

2" (51)

Plan view

6" (152)

3" (75)

Connecting strap of 2" x ¼" (50 x 6.3)

R

1½" (38) dia. hole, chamfer at
45° x 3/16" (5)

¼" (6.3)

15

The wheel rim is placed in the jig and the wheel rim adjuster bolts tightened all round to bring the wheel rim seam edges together. The rim seam joint is then clamped, the seam tack welded and the wheel rim removed from the jig to permit completion of full welding both sides down the rim seam joint.

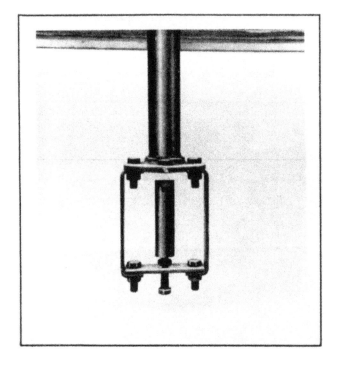

The wheel rim is replaced in the jig, and the axle is inserted down through the upper axle guide (QR) and the lower axle guide (R), the axle end resting on the axle position adjuster bolt (R) the latter then being screwed up or down as required to bring the top of the axle in line with the plane across the upper edge of the wheel rim.

Twelve wheel spokes are cut, each of ¾'' (19 mm) diameter and 13 15/16'' (354 mm) in length.

The first six spokes are laid in the spoke guides, their inner ends converging against the axle, as shown. The wheel rim adjuster bolts are used to true the rim, and the spokes are then tack welded to the wheel rim and to the axle.

The wheel rim adjuster bolts are slackened off, the wheel lifted to clear the spoke rests and turned 30°, then lowered to the jig deck. (Note: the two bolts securing the upper axle guide have their heads cut down to 1/8" (3 mm) in height to avoid their catching on the spokes when the rim is turned 30°).

The remaining six alternate wheel spokes are laid in the spoke guides with their inner ends resting on suitably dimensioned distance pieces to position their upper edges at ½" (12.5 mm) from the end of the axle, followed by tack welding to wheel rim and axle.

The wheel and axle is then removed from the jig and the full fillet welding of all the wheel spokes carried out.

This is the integral wheel and axle design in practical use (S).

See detailed diagram overleaf.

19

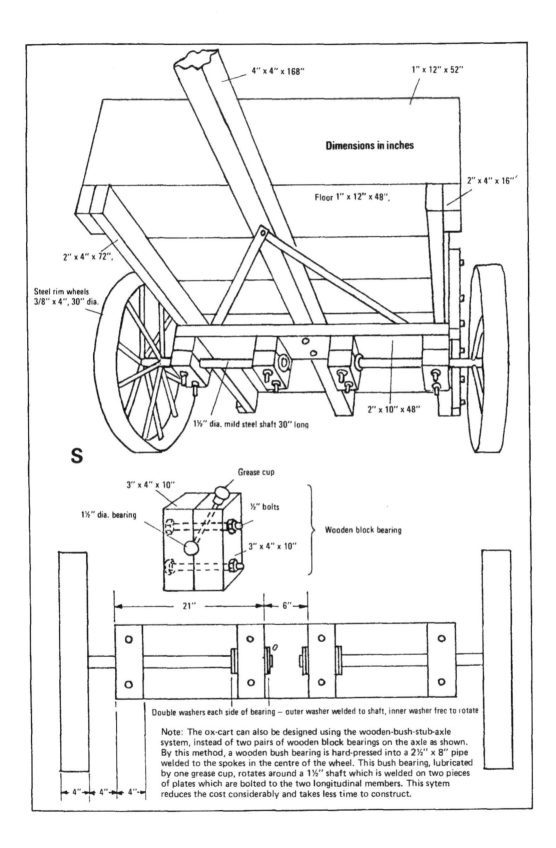

4" x 4" x 168"

1" x 12" x 52"

Dimensions in inches

2" x 4" x 16"

Floor 1" x 12" x 48",

2" x 4" x 72",

Steel rim wheels
3/8" x 4", 30" dia.

1½" dia. mild steel shaft 30" long

2" x 10" x 48"

S

Grease cup

3" x 4" x 10"

½" bolts

1½" dia. bearing

Wooden block bearing

3" x 4" x 10"

21"

6"

Double washers each side of bearing — outer washer welded to shaft, inner washer free to rotate

4" 4" 4"

Note: The ox-cart can also be designed using the wooden-bush-stub-axle
system, instead of two pairs of wooden block bearings on the axle as shown.
By this method, a wooden bush bearing is hard-pressed into a 2½" x 8" pipe
welded to the spokes in the centre of the wheel. This bush bearing, lubricated
by one grease cup, rotates around a 1½" shaft which is welded on two pieces
of plates which are bolted to the two longitudinal members. This sytem
reduces the cost considerably and takes less time to construct.

Summary

The development of a hand-operated metal-bending machine using common grade mild steel and constructed by fabrication technique has been successfully carried out, bearing in mind the aim to arrive at a design which can be made locally in developing countries and at a cost within the resources of rural craftsmen.

It should be noted that separate circle bending formers have to be built to suit each particular hoop diameter as required.

The machine is versatile in that it can be fitted up to carry out other bending operations in addition to forming hoop iron into circles, for example:

a. To bend notched angle-iron to any angle up to 90°, the bending former(L) is placed at the end of the arms with the pivot pin (G) passing through holes No. 1, two material grip/formers being used, one placed with fittings pin (N) in No. 3 hole of the fixed arm and the other placed with fittings pin (P) in No. 3 hole of the pivoting arm. The centre of the notch in the angle-iron is positioned precisely against the corner of bending former (L) and the pivoting arm moved round until the required angle is obtained.

b. In a similar operating position to that described in (a) above, mild steel flat can be bent to any sharp-cornered angle up to 90° using either the bending former (L) or a material grip/former (J) on the pivot pin (G) at holes position No. 1 in both machine arms.

Mild steel flat can also be bent to a smooth curve-cornered angle up to 180° by bending direct around the pivot pin (G) or by using the cyclindrical roller/former (K) on the pivot pin (G), using a material grip (J) in each of the machine arms.

The bending-machine design can be modified to suit the local availability of steel stock sizes, and can be further developed by addition of other types of formers and fittings to expand its usefulness to other bending operations which may be locally required according to the types of rural/agricultural tools and equipment which are to be built and maintained in the rural areas.

Milton Keynes UK
Ingram Content Group UK Ltd.
UKHW051320070824
1191UKWH00052B/394